MISSING LINK

Frank Herbert

Missing Link

Table of Contents

Epigraph
Missing Link

Epigraph

The Romantics used to say that the eyes were the windows of the Soul. A good Alien Xenologist might not put it quite so poetically ... but he can, if he's sharp, read a lot in the look of an eye!

Illustrated by van Dongen

Missing Link

"We ought to scrape this planet clean of every living thing on it," muttered Umbo Stetson, section chief of Investigation & Adjustment.

Stetson paced the landing control bridge of his scout cruiser. His footsteps grated on a floor that was the rear wall of the bridge during flight. But now the ship rested on its tail fins—all four hundred glistening red and black meters of it. The open ports of the bridge looked out on the jungle roof of Gienah III some one hundred fifty meters below. A butter yellow sun hung above the horizon, perhaps an hour from setting.

"Clean as an egg!" he barked. He paused in his round of the bridge, glared out the starboard port, spat into the fire–blackened circle that the cruiser's jets had burned from the jungle.

The I–A section chief was dark–haired, gangling, with large head and big features. He stood in his customary slouch, a stance not improved by sacklike patched blue fatigues. Although on this present operation he rated the flag of a division admiral, his fatigues carried no insignia. There was a general unkempt, straggling look about him.

Lewis Orne, junior I–A field man with a maiden diploma, stood at the opposite port, studying the jungle horizon. Now and then he glanced at the bridge control console, the chronometer above it, the big translite map of their position tilted from the opposite bulkhead. A heavy planet native, he felt vaguely uneasy on this Gienah III with its gravity of only seven–eighths Terran Standard. The surgical scars on his neck where the micro–communications equipment had been inserted itched maddeningly. He scratched.

"Hah!" said Stetson. "Politicians!"

A thin black insect with shell–like wings flew in Orne's port, settled in his close–cropped red hair. Orne pulled the insect gently from his hair, released it. Again it tried to land in his hair. He ducked. It flew across the bridge, out the port beside Stetson.

There was a thick–muscled, no–fat look to Orne, but something about his blocky, off–center features suggested a clown.

"I'm getting tired of waiting," he said.

"*You're* tired! Hah!"

A breeze rippled the tops of the green ocean below them. Here and there, red and purple flowers jutted from the verdure, bending and nodding like an attentive audience.

"Just look at that blasted jungle!" barked Stetson. "Them and their stupid orders!"

A call bell tinkled on the bridge control console. The red light above the speaker grid began blinking. Stetson shot an angry glance at it. "Yeah, Hal?"

"O.K., Stet. Orders just came through. We use Plan C. ComGO says to brief the field man, and jet out of here."

"Did you ask them about using another field man?"

Orne looked up attentively.

The speaker said: "Yes. They said we have to use Orne because of the records on the *Delphinus*."

"Well then, will they give us more time to brief him?"

"Negative. It's crash priority. ComGO expects to blast the planet anyway."

Stetson glared at the grid. "Those fat–headed, lard–bottomed, pig–brained … POLITICIANS!" He took two deep breaths, subsided. "O.K. Tell them we'll comply."

"One more thing, Stet."

"What now?"

"I've got a confirmed contact."

Instantly, Stetson was poised on the balls of his feet, alert. "Where?"

"About ten kilometers out. Section AAB–6."

"How many?"

"A mob. You want I should count them?"

"No. What're they doing?"

"Making a beeline for us. You better get a move on."

"O.K. Keep us posted."

"Right."

* * * * *

Stetson looked across at his junior field man. "Orne, if you decide you want out of this assignment, you just say the word. I'll back you to the hilt."

"Why should I want out of my first field assignment?"

"Listen, and find out." Stetson crossed to a tilt–locker behind the big translite map, hauled out a white coverall uniform with gold insignia, tossed it to Orne. "Get into these while I brief you on the map."

"But this is an R&R uni—" began Orne.

"Get that uniform on your ugly frame!"

"Yes, sir, Admiral Stetson, sir. Right away, sir. But I thought I was through with old Rediscovery & Reeducation when you drafted me off of Hamal into the I–A ... sir." He began changing from the I–A blue to the R&R white. Almost as an afterthought, he said: "...Sir."

A wolfish grin cracked Stetson's big features. "I'm soooooo happy you have the proper attitude of subservience toward authority."

Orne zipped up the coverall uniform. "Oh, yes, sir ... sir."

"O.K., Orne, pay attention." Stetson gestured at the map with its green superimposed grid squares. "Here we are. Here's that city we flew over on our way down. You'll head for it as soon as we drop you. The place is big enough that if you hold a course roughly northeast you can't miss it. We're —"

Again the call bell rang.

"What is it this time, Hal?" barked Stetson.

"They've changed to Plan H, Stet. New orders cut."

"Five days?"

"That's all they can give us. ComGO says he can't keep the information out of High Commissioner Bullone's hands any longer than that."

"It's five days for sure then."

"Is this the usual R&R foul–up?" asked Orne.

Stetson nodded. "Thanks to Bullone and company! We're just one jump ahead of catastrophe, but they still pump the bushwah into the Rah & Rah boys back at dear old Uni–Galacta!"

"You're making light of my revered alma mater," said Orne. He struck a pose. "We must reunite the lost planets with our centers of culture and industry, and take up the *glor*–ious onward march of mankind that was so *bru*–tally—"

"Can it!" snapped Stetson. "We both know we're going to rediscover one planet too many some day. Rim War all over again. But this is a different breed of fish. It's not, repeat, *not* a *re*–discovery."

Orne sobered. "Alien?"

"Yes. A–L–I–E–N! A never–before–contacted culture. That language you were force fed on the way over, that's an alien language. It's not complete … all we have off the *minis*. And we excluded data on the natives because we've been hoping to dump this project and nobody the wiser."

"Holy mazoo!"

"Twenty–six days ago an I–A search ship came through here, had a routine mini–sneaker look at the place. When he combed in his net of sneakers to check the tapes and films, lo and behold, he had a little stranger."

"One of *theirs*?"

"No. It was a *mini* off the *Delphinus Rediscovery*. The *Delphinus* has been unreported for eighteen standard months!"

"Did it crack up here?"

"We don't know. If it did, we haven't been able to spot it. She was supposed to be way off in the Balandine System by now. But we've something else on our minds. It's the one item that makes me want to blot out this place, and run home with my tail between my legs. We've a—"

Again the call bell chimed.

"NOW WHAT?" roared Stetson into the speaker.

"I've got a *mini* over that mob, Stet. They're talking about us. It's a definite raiding party."

"What armament?"

"Too gloomy in that jungle to be sure. The infra beam's out on this *mini*. Looks like hard pellet rifles of some kind. Might even be off the *Delphinus*."

"Can't you get closer?"

"Wouldn't do any good. No light down there, and they're moving up fast."

"Keep an eye on them, but don't ignore the other sectors," said Stetson.

"You think I was born yesterday?" barked the voice from the grid. The contact broke off with an angry sound.

* * * * *

"One thing I like about the I–A," said Stetson. "It collects such even–tempered types." He looked at the white uniform on Orne, wiped a hand across his mouth as though he'd tasted something dirty.

"Why *am* I wearing this thing?" asked Orne.

"Disguise."

"But there's no mustache!"

Stetson smiled without humor. "That's one of I–A's answers to those fat–keistered politicians. We're setting up our own search system to find the planets before *they* do. We've managed to put spies in key places at R&R. Any touchy planets our spies report, we divert the files."

"Then what?"

"Then we look into them with bright boys like you—disguised as R&R field men."

"Goody, goody. And what happens if R&R stumbles onto me while I'm down there playing patty cake?"

"We disown you."

"But you said an I–A ship found this joint."

"It did. And then one of our spies in R&R intercepted a *routine* request for an agent–instructor to be assigned here with full equipment. Request signed by a First–Contact officer name of Diston ... of the *Delphinus*!"

"But the Del—"

"Yeah. Missing. The request was a forgery. Now you see why I'm mostly for rubbing out this place. Who'd dare forge such a thing unless he knew for sure that the original FC officer was missing ... or dead?"

"What the jumped up mazoo are we doing here, Stet?" asked Orne. "Alien calls for a full contact team with all of the—"

"It calls for one planet–buster bomb ... buster—in five days. Unless you give them a white bill in the meantime. High Commissioner Bullone will have word of this planet by then. If Gienah III still exists in five days, can't you imagine the fun the politicians'll have with it? Mama mia! We want this planet cleared for contact or dead before then."

"I don't like this, Stet."

"YOU don't like it!"

"Look," said Orne. "There must be another way. Why ... when we teamed up with the Alerinoids we gained five hundred years in the physical sciences alone, not to mention the—"

"The Alerinoids didn't knock over one of our survey ships first."

"What if the *Delphinus* just crashed here ... and the locals picked up the pieces?"

"That's what you're going in to find out, Orne. But answer me this: If they *do* have the *Delphinus*, how long before a tool–using race could be a threat to the galaxy?"

"I saw that city they built, Stet. They could be dug in within six months, and there'd be no—"

"Yeah."

Orne shook his head. "But think of it: Two civilizations that matured along different lines! Think of all the different ways we'd approach the same problems ... the lever that'd give us for—"

"You sound like a Uni–Galacta lecture! Are you through marching arm in arm into the misty future?"

Orne took a deep breath. "Why's a freshman like me being tossed into this dish?"

"You'd still be on the *Delphinus* master lists as an R&R field man. That's important if you're masquerading."

"Am I the only one? I know I'm a recent *convert*, but—"

"You want out?"

"I didn't say that. I just want to know why I'm—"

"Because the bigdomes fed a set of requirements into one of their iron monsters. Your card popped out. They were looking for somebody capable, dependable ... and ... *expendable!*"

"Hey!"

"That's why I'm down here briefing you instead of sitting back on a flagship. *I* got you into the I–A. Now, you listen carefully: If you push the panic button on this one without cause, I will personally flay you alive. We both know the advantages of an alien contact. But if you get into a hot spot, and call for help, I'll dive this cruiser into that city to get you out!"

Orne swallowed. "Thanks, Stet. I'm—"

* * * * *

"We're going to take up a tight orbit. Out beyond us will be five transports full of I–A marines and a Class IX Monitor with one planet–buster. You're calling the shots, God help you! First, we want to know if they have the *Delphinus* ... and if so, where it is. Next, we want to know just how warlike these goons are. Can we control them if they're bloodthirsty. What's their potential?"

"In five days?"

"Not a second more."

"What do we know about them?"

"Not much. They look something like an ancient Terran chimpanzee ... only with blue fur. Face is hairless, pink–skinned." Stetson snapped a switch. The translite map became a screen with a figure frozen on it. "Like that. This is life size."

"Looks like the missing link they're always hunting for," said Orne. "Yeah, but you've got a different kind of a missing link."

"Vertical–slit pupils in their eyes," said Orne. He studied the figure. It had been caught from the front by a mini–sneaker camera. About five feet tall. The stance was slightly bent forward, long arms. Two vertical nose slits. A flat, lipless mouth. Receding chin. Four–fingered hands. It wore a wide belt from which dangled neat pouches and what looked like tools, although their use was obscure. There appeared to be the tip of a tail protruding from behind one of the squat legs. Behind the creature towered the faery spires of the city they'd observed from the air.

"Tails?" asked Orne.

"Yeah. They're arboreal. Not a road on the whole planet that we can find. But there are lots of vine lanes through the jungles." Stetson's face hardened. "Match *that* with a city as advanced as that one."

"Slave culture?"

"Probably."

"How many cities have they?"

"We've found two. This one and another on the other side of the planet. But the other one's a ruin."

"A ruin? Why?"

"You tell us. Lots of mysteries here."

"What's the planet like?"

"Mostly jungle. There are polar oceans, lakes and rivers. One low mountain chain follows the equatorial belt about two thirds around the planet."

"But only two cities. Are you sure?"

"Reasonably so. It'd be pretty hard to miss something the size of that thing we flew over. It must be fifty kilometers long and at least ten wide. Swarming with these creatures, too. We've got a zone–count estimate that places the city's population at over thirty million."

"Whee–ew! Those are tall buildings, too."

"We don't know much about this place, Orne. And unless you bring them into the fold, there'll be nothing but ashes for our archaeologists to pick over."

"Seems a dirty shame."

"I agree, but—"

The call bell jangled.

* * * * *

Stetson's voice sounded tired: "Yeah, Hal?"

"That mob's only about five kilometers out, Stet. We've got Orne's gear outside in the disguised air sled."

"We'll be right down."

"Why a disguised sled?" asked Orne.

"If they think it's a ground buggy, they might get careless when you most need an advantage. We could always scoop you out of the air, you know."

"What're my chances on this one, Stet?"

Stetson shrugged. "I'm afraid they're slim. These goons probably have the *Delphinus*, and they want you just long enough to get your equipment and everything you know."

"Rough as that, eh?"

"According to our best guess. If you're not out in five days, we blast."

Orne cleared his throat.

"Want out?" asked Stetson.

"No."

"Use the *back–door* rule, son. Always leave yourself a way out. Now … let's check that equipment the surgeons put in your neck." Stetson put a hand to his throat. His mouth remained closed, but there was a surf–hissing voice in Orne's ears: "You read me?"

"Sure. I can—"

"No!" hissed the voice. "Touch the mike contact. Keep your mouth closed. Just use your speaking muscles without speaking."

Orne obeyed.

"O.K.," said Stetson. "You come in loud and clear."

"I ought to. I'm right on top of you!"

"There'll be a relay ship over you all the time," said Stetson. "Now ... when you're not touching that mike contact this rig'll still feed us what you say ... and everything that goes on around you, too. We'll monitor everything. Got that?"

"Yes."

Stetson held out his right hand. "Good luck. I meant that about diving in for you. Just say the word."

"I know the word, too," said Orne. "HELP!"

* * * * *

Gray mud floor and gloomy aisles between monstrous bluish tree trunks—that was the jungle. Only the barest weak glimmering of sunlight penetrated to the mud. The disguised sled—its para–grav units turned off—lurched and skidded around buttress roots. Its headlights swung in wild arcs across the trunks and down to the mud. Aerial creepers—great looping vines of them—swung down from the towering forest ceiling. A steady drip of condensation spattered the windshield, forcing Orne to use the wipers.

In the bucket seat of the sled's cab, Orne fought the controls. He was plagued by the vague slow–motion–floating sensation that a heavy planet native always feels in lighter gravity. It gave him an unhappy stomach.

Things skipped through the air around the lurching vehicle: flitting and darting things. Insects came in twin cones, siphoned toward the headlights. There was an endless chittering whistling tok–tok–toking in the gloom beyond the lights.

Stetson's voice hissed suddenly through the surgically implanted speaker:

"How's it look?"

"Alien."

"Any sign of that mob?"

"Negative."

"O.K. We're taking off."

Behind Orne, there came a deep rumbling roar that receded as the scout cruiser climbed its jets. All other sounds hung suspended in after–silence, then resumed: the strongest first and then the weakest.

A heavy object suddenly arced through the headlights, swinging on a vine. It disappeared behind a tree. Another. Another. Ghostly shadows with vine pendulums on both sides. Something banged down heavily onto the hood of the sled.

Orne braked to a creaking stop that shifted the load behind him, found himself staring through the windshield at a native of Gienah III. The native crouched on the hood, a Mark XX exploding–pellet rifle in his right hand directed at Orne's head. In the abrupt shock of meeting, Orne recognized the weapon: standard issue to the marine guards on all R&R survey ships.

The native appeared the twin of the one Orne had seen on the translite screen. The four–fingered hand looked extremely capable around the stock of the Mark XX.

Slowly, Orne put a hand to his throat, pressed the contact button. He moved his speaking muscles: *"Just made contact with the mob. One on the hood now has one of our Mark XX rifles aimed at my head."*

The surf–hissing of Stetson's voice came through the hidden speaker: *"Want us to come back?"*

"Negative. Stand by. He looks cautious rather than hostile."

Orne held up his right hand, palm out. He had a second thought: held up his left hand, too. Universal symbol of peaceful intentions: empty hands. The gun muzzle lowered slightly. Orne called into his mind the language that had been hypnoforced into him. *Ocheero? No. That means 'The People.' Ah* ... And he had the heavy fricative greeting sound.

"Ffroiragrazzi," he said.

The native shifted to the left, answered in pure, unaccented High Galactese: "Who are you?"

Orne fought down a sudden panic. The lipless mouth had looked so odd forming the familiar words.

Stetson's voice hissed: *"Is that the native speaking Galactese?"*

Orne touched his throat. *"You heard him."*

He dropped his hand, said: "I am Lewis Orne of Rediscovery and Reeducation. I was sent here at the request of the First–Contact officer on the *Delphinus Rediscovery*."

"Where is your ship?" demanded the Gienahn.

"It put me down and left."

"Why?"

"It was behind schedule for another appointment."

* * * * *

Out of the corners of his eyes, Orne saw more shadows dropping to the mud around him. The sled shifted as someone climbed onto the load behind the cab. The someone scuttled agilely for a moment.

The native climbed down to the cab's side step, opened the door. The rifle was held at the ready. Again, the lipless mouth formed Galactese words: "What do you carry in this ... vehicle?"

"The equipment every R&R field man uses to help the people of a rediscovered planet improve themselves." Orne nodded at the rifle. "Would you mind pointing that weapon some other direction? It makes me nervous."

The gun muzzle remained unwaveringly on Orne's middle. The native's mouth opened, revealing long canines. "Do we not look strange to you?"

"I take it there's been a heavy mutational variation in the humanoid norm on this planet," said Orne. "What is it? Hard radiation?"

No answer.

"It doesn't really make any difference, of course," said Orne. "I'm here to help you."

"I am Tanub, High Path Chief of the Grazzi," said the native. "I decide who is to help."

Orne swallowed.

"Where do you go?" demanded Tanub.

"I was hoping to go to your city. Is it permitted?"

A long pause while the vertical–slit pupils of Tanub's eyes expanded and contracted. "It is permitted."

Stetson's voice came through the hidden speaker: *"All bets off. We're coming in after you. That Mark XX is the final straw. It means they have the* Delphinus *for sure!"*

Orne touched his throat. *"No! Give me a little more time!"*

"Why?"

"I have a hunch about these creatures."

"What is it?"

"No time now. Trust me."

Another long pause in which Orne and Tanub continued to study each other. Presently, Stetson said: *"O.K. Go ahead as planned. But find out where the* Delphinus *is! If we get that back we pull their teeth."*

"Why do you keep touching your throat?" demanded Tanub.

"I'm nervous," said Orne. "Guns always make me nervous."

The muzzle lowered slightly.

"Shall we continue on to your city?" asked Orne. He wet his lips with his tongue. The cab light on Tanub's face was giving the Gienahn an eerie sinister look.

"We can go soon," said Tanub.

"Will you join me inside here?" asked Orne. "There's a passenger seat right behind me."

Tanub's eyes moved catlike: right, left. "Yes." He turned, barked an order into the jungle gloom, then climbed in behind Orne.

"When do we go?" asked Orne.

"The great sun will be down soon," said Tanub. "We can continue as soon as Chiranachuruso rises."

"Chiranachuruso?"

"Our satellite … our moon," said Tanub.

"It's a beautiful word," said Orne. "Chiranachuruso."

"In our tongue it means: The Limb of Victory," said Tanub. "By its light we will continue."

Orne turned, looked back at Tanub. "Do you mean to tell me that you can see by what light gets down here through those trees?"

"Can you not see?" asked Tanub.

"Not without the headlights."

"Our eyes differ," said Tanub. He bent toward Orne, peered. The vertical slit pupils of his eyes expanded, contracted. "You are the same as the … others."

"Oh, on the *Delphinus*?"

Pause. "Yes."

Presently, a greater gloom came over the jungle, bringing a sudden stillness to the wild life. There was a chittering commotion from the natives in the trees around the sled. Tanub shifted behind Orne.

"We may go now," he said. "Slowly … to stay behind my … scouts."

"Right." Orne eased the sled forward around an obstructing root.

* * * * *

Silence while they crawled ahead. Around them shapes flung themselves from vine to vine.

"I admired your city from the air," said Orne. "It is very beautiful."

"Yes," said Tanub. "Why did you land so far from it?"

"We didn't want to come down where we might destroy anything."

"There is nothing to destroy in the jungle," said Tanub.

"Why do you have such a big city?" asked Orne.

Silence.

"I said: Why do you—"

"You are ignorant of our ways," said Tanub. "Therefore, I forgive you. The city is for our race. We must breed and be born in sunlight. Once—long ago—we used crude platforms on the tops of the trees. Now …only the … wild ones do this."

Stetson's voice hissed in Orne's ears: *"Easy on the sex line, boy. That's always touchy. These creatures are oviparous. Sex glands are apparently hidden in that long fur behind where their chins ought to be."*

"Who controls the breeding sites controls our world," said Tanub. "Once there was another city. We destroyed it."

"Are there many … wild ones?" asked Orne.

"Fewer each year," said Tanub.

"There's how they get their slaves," hissed Stetson.

"You speak excellent Galactese," said Orne.

"The High Path Chief commanded the best teacher," said Tanub. "Do you, too, know many things, Orne?"

"That's why I was sent here," said Orne.

"Are there many planets to teach?" asked Tanub.

"Very many," said Orne. "Your city—I saw very tall buildings. Of what do you build them?"

"In your tongue—glass," said Tanub. "The engineers of the *Delphinus* said it was impossible. As you saw—they are wrong."

"A glass–blowing culture," hissed Stetson. *"That'd explain a lot of things."*

Slowly, the disguised sled crept through the jungle. Once, a scout swooped down into the headlights, waved. Orne stopped on Tanub's order, and they waited almost ten minutes before proceeding.

"Wild ones?" asked Orne.

"Perhaps," said Tanub.

A glowing of many lights grew visible through the giant tree trunks. It grew brighter as the sled crept through the last of the jungle, emerged in cleared land at the edge of the city.

Orne stared upward in awe. The city fluted and spiraled into the moonlit sky. It was a fragile appearing lacery of bridges, winking dots of light. The bridges wove back and forth from building to building until the entire visible network appeared one gigantic dew–glittering web.

"All that with glass," murmured Orne.

"What's happening?" hissed Stetson.

Orne touched his throat contact. *"We're just into the city clearing, proceeding toward the nearest building."*

"This is far enough," said Tanub.

* * * * *

Orne stopped the sled. In the moonlight, he could see armed Gienahns all around. The buttressed pedestal of one of the buildings loomed directly ahead. It looked taller than had the scout cruiser in its jungle landing circle.

Tanub leaned close to Orne's shoulder. "We have not deceived you, have we, Orne?"

"Huh? What do you mean?"

"You have recognized that we are not mutated members of your race."

Orne swallowed. Into his ears came Stetson's voice: *"Better admit it."*

"That's true," said Orne.

"I like you, Orne," said Tanub. "You shall be one of my slaves. You will teach me many things."

"How did you capture the *Delphinus*?" asked Orne.

"You know that, too?"

"You have one of their rifles," said Orne.

"Your race is no match for us, Orne … in cunning, in strength, in the prowess of the mind. Your ship landed to repair its tubes. Very inferior ceramics in those tubes."

Orne turned, looked at Tanub in the dim glow of the cab light. "Have you heard about the I–A, Tanub?"

"I–A? What is that?" There was a wary tenseness in the Gienahn's figure. His mouth opened to reveal the long canines.

"You took the *Delphinus* by treachery?" asked Orne.

"They were simple fools," said Tanub. "We are smaller, thus they thought us weaker." The Mark XX's muzzle came around to center on Orne's stomach. "You have not answered my question. What is the I–A?"

"I am of the I–A," said Orne. "Where've you hidden the *Delphinus*?"

"In the place that suits us best," said Tanub. "In all our history there has never been a better place."

"What do you plan to do with it?" asked Orne.

"Within a year we will have a copy with our own improvements. After that —"

"You intend to start a war?" asked Orne.

"In the jungle the strong slay the weak until only the strong remain," said Tanub.

"And then the strong prey upon each other?" asked Orne.

"That is a quibble for women," said Tanub.

"It's too bad you feel that way," said Orne. "When two cultures meet like this they tend to help each other. What have you done with the crew of the *Delphinus*?"

"They are slaves," said Tanub. "Those who still live. Some resisted. Others objected to teaching us what we want to know." He waved the gun muzzle. "You will not be that foolish, will you, Orne?"

"No need to be," said Orne. "I've another little lesson to teach you: I already know where you've hidden the *Delphinus*."

"Go, boy!" hissed Stetson. *"Where is it?"*

"Impossible!" barked Tanub.

"It's on your moon," said Orne. "Darkside. It's on a mountain on the darkside of your moon."

Tanub's eyes dilated, contracted. "You read minds?"

"The I–A has no need to read minds," said Orne. "We rely on superior mental prowess."

"The marines are on their way," hissed Stetson. *"We're coming in to get you. I'm going to want to know how you guessed that one."*

"You are a weak fool like the others," gritted Tanub.

"It's too bad you formed your opinion of us by observing only the low grades of the R&R," said Orne.

"Easy, boy," hissed Stetson. *"Don't pick a fight with him now. Remember, his race is arboreal. He's probably as strong as an ape."*

"I could kill you where you sit!" grated Tanub.

"You write finish for your entire planet if you do," said Orne. "I'm not alone. There are others listening to every word we say. There's a ship overhead that could split open your planet with one bomb—wash it with molten rock. It'd run like the glass you use for your buildings."

"You are lying!"

"We'll make you an offer," said Orne. "We don't really want to exterminate you. We'll give you limited membership in the Galactic Federation until you prove you're no menace to us."

"Keep talking," hissed Stetson. *"Keep him interested."*

"You dare insult me!" growled Tanub.

"You had better believe me," said Orne. "We—"

Stetson's voice interrupted him: *"Got it, Orne! They caught the* Delphinus *on the ground right where you said it'd be! Blew the tubes off it. Marines now mopping up."*

"It's like this," said Orne. "We already have recaptured the *Delphinus*." Tanub's eyes went instinctively skyward. "Except for the captured armament you still hold, you obviously don't have the weapons to meet us," continued Orne. "Otherwise, you wouldn't be carrying that rifle off the *Delphinus*."

"If you speak the truth, then we shall die bravely," said Tanub.

"No need for you to die," said Orne.

"Better to die than be slaves," said Tanub.

"We don't need slaves," said Orne. "We—"

"I cannot take the chance that you are lying," said Tanub. "I must kill you now."

* * * * *

Orne's foot rested on the air sled control pedal. He depressed it. Instantly, the sled shot skyward, heavy G's pressing them down into the seats. The gun in Tanub's hands was slammed into his lap. He struggled to raise it. To Orne, the weight was still only about twice that of his home planet of Chargon. He reached over, took the rifle, found safety belts, bound Tanub with them. Then he eased off the acceleration.

"We don't need slaves," said Orne. "We have machines to do our work. We'll send experts in here, teach you people how to exploit your planet, how to build good transportation facilities, show you how to mine your minerals, how to—"

"And what do we do in return?" whispered Tanub.

"You could start by teaching us how you make superior glass," said Orne. "I certainly hope you see things our way. We really don't want to have to come down there and clean you out. It'd be a shame to have to blast that city into little pieces."

Tanub wilted. Presently, he said: "Send me back. I will discuss this with … our council." He stared at Orne. "You I–A's are too strong. We did not know."

* * * * *

In the wardroom of Stetson's scout cruiser, the lights were low, the leather chairs comfortable, the green beige table set with a decanter of Hochar brandy and two glasses.

Orne lifted his glass, sipped the liquor, smacked his lips. "For a while there, I thought I'd never be tasting anything like this again."

Stetson took his own glass. "ComGO heard the whole thing over the general monitor net," he said. "D'you know you've been breveted to senior field man?"

"Ah, they've already recognized my sterling worth," said Orne.

The wolfish grin took over Stetson's big features. "Senior field men last about half as long as the juniors," he said. "Mortality's terrific?"

"I might've known," said Orne. He took another sip of the brandy.

Stetson flicked on the switch of a recorder beside him. "O.K. You can go ahead any time."

"Where do you want me to start?"

"First, how'd you spot right away where they'd hidden the *Delphinus*?"

"Easy. Tanub's word for his people was *Grazzi*. Most races call themselves something meaning *The People*. But in his tongue that's *Ocheero*. *Grazzi* wasn't on the translated list. I started working on it. The most likely answer was that it had been adopted from another language, and meant *enemy*."

"And *that* told you where the *Delphinus* was?"

"No. But it fitted my hunch about these Gienahns. I'd kind of felt from the first minute of meeting them that they had a culture like the Indians of ancient Terra."

"Why?"

"They came in like a primitive raiding party. The leader dropped right onto the hood of my sled. An act of bravery, no less. Counting coup, you see?"

"I guess so."

"Then he said he was High Path Chief. That wasn't on the language list, either. But it was easy: *Raider Chief*. There's a word in almost every language in history that means raider and derives from a word for road, path or highway."

"Highwaymen," said Stetson.

"Raid itself," said Orne. "An ancient Terran language corruption of road."

"Yeah, yeah. But where'd all this translation griff put—"

"Don't be impatient. Glass–blowing culture meant they were just out of the primitive stage. That, we could control. Next, he said their moon was *Chiranachuruso*, translated as *The Limb of Victory*. After that it just fell into place."

"How?"

"The vertical–slit pupils of their eyes. Doesn't that mean anything to you?"

"Maybe. What's it mean to you?"

"Night–hunting predator accustomed to dropping upon its victims from above. No other type of creature ever has had the vertical slit. And Tanub said himself that the *Delphinus* was hidden in the best place in all of their history. History? That'd be a high place. Dark, likewise. Ergo: a high place on the darkside of their moon."

"I'm a pie–eyed greepus," whispered Stetson.

Orne grinned, said: "You probably are … sir."

THE END

www.ingramcontent.com/pod-product-compliance
Lightning Source LLC
Chambersburg PA
CBHW030046230526
45472CB00005B/1697